John [illegible]

John Moritz

cartography

Selected Poems
1968 ~ 1998

First Intensity Press
Lawrence, Kansas - 2001

Some of these poems, often in different versions, first appeared in the following magazines and books, or as broadsides: *Truck* (David Wilk, ed.), *Tansy* (John Moritz, ed.), *Cottonwood Review* (Jim Schmidt, ed.), *30 Kansas Poets* (Denise Low, ed.), *Man Root* (Paul Maria, ed.), *Glitch* (Paul McDonough and Jane McGriff, eds.), *Bezoar* (Paul Kahn, ed.), *Io* (Richard Grossinger, ed.), *Coronado Press* (John Longhurst), *Aux-Arcs* (Robert Greene, ed.), *Zelot* (Richard Blevins, ed.), *Friends of K.U. Poetry Collection* (Robert Melton), *:that:* (Stephen Ellis and Steven Dignazio, eds.), *Avec* (Cydney Chadwick, ed., Jim McCrary, contrib. ed.), *First Intensity* (Lee Chapman, ed.), *Crossings* (Tansy Press, 1969), *Consentryks* (Zelot-Tansy Press, 1985).

~ ★ ~

ISBN 1-889960-02-0

First Edition

Published by
First Intensity Press
P.O. Box 665
Lawrence, Kansas 66044

Book design & typography by Lee Chapman

~

I want to relate the respect I have for Bill Rodgers and Roger Christie who would have been very much a part of this book's production if things had worked out differently. My gratitude, once again, goes to Curtis Dillon fort his help proofreading what was a difficult text. Sharon, thank you for putting up with my rant. Finally, I want to bless dear Lee Chapman who was there from the beginning with the cover for *Crossings* and who has illustrated my work ever since with the intuitive powers of an alchemist. I can't express enough how much she is a part of it.

~

Cartography

Night falls over abandoned hills, over the brown of grass,
over those places the heart dwells. Those dark waves of
[inland sea
pitch on the arm of this vast loneliness.
The approach opens in the west. The map of a day gone
across the toss of bluestem hangs over the far slope.
A few scrub cottonwood, hedging the memory of a creek bed,
frame a resistance. In the last length of an orange
[filtered light,
the filter of sandstone passes through the broken tangle of
[branch.

Table of Contents

Photographs & Illustrations

Crossings I - V

I

the river was
just a slow
trickle that August
in 67
        crossing the bridge
where the land falls
along the willow banks

& the town be Lawrence.

her namesake
            Bostonian
of the emigrant aid
sent sharps dressed as bibles
for the resistance.

one Dr. Charles Robinson
with 29 others
from Massachusetts
made the crossing.

the site:
    lat 38.58, long 95.14
the date:
    August 1, 1854
marked—Free State

Later Dr. Robinson became the states first governor.

at the ford
Sara Robinson observed the Kansa
river Kansa, desperate
& new to contact
abandoned their maize,
                    selling trinkets
                       at the crotch of Oregon
                                   & Santa Fe

the billboard of acculturation
is the neon over Jackys.
I sit watching the cowboy swagger
a Kickapoo drinking beer
until he falls & somebody
telephones a cab
to send him back to Haskell
where he tells me he studies electronics
plans to move to Oklahoma
& fix tvs for his OK brothers
the Cherokee, *cadillac indians*
whose mythic oil wells
say well worth the walk from Georgia
his name is Frank
& insists I buy him a beer

                        did you buy their beads?
                        did you buy their blankets, Sara?

Dr. Robinson, the Gandhi of bleeding Kansas
built on Oread, "Good place for a university"

Sara, facing the sunset
remembered the documents
of New England, imagined
                                        the Pacific
                                                              merged
on a rolling wave of brome.

that field
                 her vision
climaxed by the ladies
at the college
                         social teas
                         appleseed
                         tree planting
                                              affairs
       oak   maple
                                   even evergreen
break waves on the landscape.

II

bottomlander seeks high ground above the river
Monchonsia with a song in his heart

1844
        he was defined briefly
"cornbeansquash"
not lasting area as long
as buffalo.
                Monchonsia
Kansa, danct on this promontory
his songs were oroborian
earth   or limestone chip strata
above the two flooded waters
                                Kansas & Wakarusa
                                                rising.

the bag I see floating
holds no secrets, is not filled
with visions & songs
well maybe a song
but common
parfleche split open
pemmican, a tobacco leaf or two
maybe some green feathers
swept downstream
maybe a river song

III

> "The whole question of slavery
> depends upon the hemp crop."
> —*Senator James Lane*

so the point is Lawrence
kid the weed dries in attics
root up & read the gold leaf
on the Kaw Valley Hemp
Pickers Association
matchbook & strike up the band
to match that, Ole John went
to Harper's & Lane took
the RR with him everywhere
went to congress & Robinson
sold his land
to the University

but the kids cut it
in August, coins spread
the Outlaw
snapshots give us back
young Billy
grinning buck tooth
"easy as shootin' ground
hog"

while packaging a bundle
for the coast.

from *Myths & Legends of Flowers, Trees, Fruits & Plants*

> "I sow the hemp seed
> Hemp seed I sow
> He that loves me best
> Come after me now."

churchyard maiden
back fence, yes
with a moon
to be called upon
she steps a dizzy rumba
her weed legs
know the wind
she's a courtin'

enter:
a man
with a scythe
swinging
at a phantom
crop
that springs
from her footsteps

IV

& from an old world
emerged a map
Nuevo Mundi
                    with a place
called Quivira
                    by some
which furnished the Spanish
their trip to the new world
for the loot

                    can't separate
man from mount
mouth from iron clad
these were satyrs
who usurpt with godlike proportions.

what mysteries ride these winds?
what is the taste of gold burning the eyes?

But there was
Cabeza de Vaca
& Estevan, the moor
first of Europa, shipwrecked
following the cow trails
found body/spirit
fused by the spacial arc
at the vastness whole to man
more than metal was this solid
(one act) soldering
the metal man, transmutation glowing
healer & by the nature of, like Estevan
healed others
                    but the other story
is all too familiar
Coronado found no Quivira
only a land worthy for grapes
wrote the king asking
permission to return to New Spain.

at Topeka State
Historical Society an iron blade
rusted (uncovered SW Kansas corner
said to locate the end of that vision.

buffalo shit stacked
waist hi
re/turning

Donde esta la ciudad de oro?

V

thots one November morning when the fork
discovering the layer (a six inches deep
to be warm & humid.

the truck dumped silage, stopped
beyond the pit, lunch
                                        a newcomer
                                        I jumped the clutch
tractor lunged in gear, winter packing
should be fresh beneath the mulch, she brought
a bag of meatloaf sandwiches
                                                    her face gathered
tite at the bone by dark lines
                                            solemnity
                                                            I guessed
caution but by the retelling, desperate

as the radio rain percentage stacked
thick from the west, row after
row the double row cutter droned
                                          from the deck
postcards she collected, one a buffalo: Black
Hills, SD, the tarot

card dealt off the barb wire stretched
that morning on the dew. 180 acres
& the beast

                        defined by the corral in his lungs
snorting for air hung
on fence posts.

marble
the looking back at
(capsuled
I sit on the stump
that split a stone marked
Mild, dead of a common
date. farmers/shop
keepers awoke to the dawn of their ending
dying, the ground turned
commonly. Wisconsin boys
a stone row, shot running
toward the arsenal
Mild Ackerman
riven by a force
that once growd
the date:
August 23, 1863
20 years
Quantrill raid
all istory
ah! this skeleton, the tale made old
as the substance flows
consumed
to rot, a war at my own time
same to be shot running
looking for a jimson patch.
I sip chianti to keep this
chill away
Mild.

  & from the porch
before moon, the reminder
of what had been laid
across this prairie, were peacocks
brought by the landlords brother
as a souvenir
                        in the tree juxtaposed
with scorpius (lower claw gone in the branches.
I said
          they look like vultures, I stood
taking the lesson at the extreme,
                 one step to gold the SOPHIC
HYDROLITH . . . first red (mars), then white,
green, yellow like a peacocks tail
                              tale of what passed
on this most back porch
on this most back land.
the lite collecting moths
cri/ckets beyond the radius
                                    we drank
home brew that was bitter
too much yeast was used.

both tan
                Herb worked the hay
hauling & baling
cleaning the pens
feeding the pigs
                         & Janis
gave Margaret the recipe
for the bread she baked
said she'd be baking every
day even in the heat
if it weren't for her job in town.
                              the birds spread the fantasy
like vultures
                    their feathers fall
at equinox, the eyes sold
at a psychedelic store in town

"The state of nature as it is commonly accepted is non-existent. When man first appeared upon the scene he destroyed such a state, because he possessed the unique capacity to act with purpose."

—James C. Malin, *The Grassland of North America*

for the Carolina Parakeet

"The Parakeet does not satisfy itself with cockle-burs
but eats or destroys almost every kind
of fruit indiscriminately."
—John James Audubon
Book of American Birds

1

Audubon's rifle misfired on his first
day out, on the second
his luck was
much improved
he shot:

17 Parakeets
1 Yellow Chat
2 White-throated Finches
1 Yellow-rump Warbler
among sundry others

but sensed to paint the green ones
eating cockle-burs.

"Our attention was attracted by the
loud screaming of green birds."
—Heinrich Lienhard
Saint Louis to Sutter's Fort

2

The small are unfelt, the fast
unseen & the nocturnal
unnoticed
but the screamers are
always
the first
& by time of settlement
the Parakeet was
extinct.

O my, Caro-
lina
so fierce
for wood
couldn't
match ours
there goes my love with
green feathers
in her hair.

Coffee Conversation

why you got no eyes   johnny
                    she askd

got em shot through   baby
                    he sd

to yellow haird girl
he was thinking of
by the grill
          coffee
hamburger on formica

got em shot through?

FLO (the name tag sd)

yes pass me the ketchup
my gal FLO
          & ill sell you the
                    sunrise

how can you do that   johnny
you got no eyes?

that was the price i had to pay
          sweet thing

taking a bite
sauce still on his lips
while FLO shuffled up two cups
for truckers at booth
she brushd back yellow haird
behind the counter again
wiping the formica

FLO (the name tag

alright what do you want for it?
she askd

your eyes

for Max

the *air age* slaps the brush below the tri-blades
of a Hughes two seater, flushing a coyote out

onto the open face of plain

a carbine snaps.

Rosalea's Hotel, Harper, Kansas 1968

a note for the girlz & boyz passing through on their way to
sunne California.

At Webb's 98th meridian, the gist
remains hostile, the air is
stifling
up &
down the
——————— line. No shade,
they say, nor much to snort
except an ounce of Ozone
at market after an occasional
tornado sweepin' through
sleepy towns
on the express to Oz.

So when a place comes up make note.

The great wheat shag carpet seen all day
driving now rests in the lobby
reflected off foil walls. Rosalea, like a fine whisp
of prairie grass bends over to move the shuttle through the
loom.

you grow familiar with the room.

There's no marquee to mark
but red brick acts for that
past the line on an edge of a way
folks think, linear, linear
like leaning on a bank at a glance, flat
unaltered from the way
they grasp the land.

She asks if we'd seen the *hippeez*
picture in the Newton paper
getting his head shaved
& did I catch the size of the Marshal's knee?

no, but at the diner I overheard
hunting coyote from a tri-blade
though I actually hadn't, vagrant
was meant, to come down
like a machine gun not so metaphorically.

Coffee anyone? she asks & I dash
upstairs for cognac, those Mennonites
think I'm getting messages from Chairman Mao
so I left my little red book like under a rock

just then her friend Bill slides in from Colorado
sits down & plays some barrel house on the upright
& for the first time a breeze whips through.

The connection with the day was Rosalea
and her passing a shuttle through the loom of evening.

Hey refugees, next stop Boulder!

Attica, Kansas

overheard at an auction — two ladies talking
I took to be mother and daughter
the mother inquiring about her daughter's husband
who had been bitten by a rattlesnake
while in the fields harvesting wheat
and had been close to death for some time
but now seemed recovering
"It's the pesticides . . ."
the younger woman said,
"passed on in the venom"

snake   I thought I heard snake   in the quickened sufi
movement of the hands   shielding   stoned and stumbled
down the steps under   the four chained   guardian dogs
iron spiked collars   and bleeding heart   of the church
[ceiling
the movement of waist and hands I heard not snake
but wasps in the juniper   and Christopher made a silver
[amulet
to close the fear   a similarity to the skeletal snakes
of an Australian bark painting

Capt. Bly

recites frm
a serape
inciting riots
then sails away
with as much of a
vested interest
in the skull ring
as Westmoreland

both views
are done
w/ binoculars
& a hug

"It's this way. Here's this country with what accumulation it has—so many people having lived here millennia. Which ought to mean (people being active, more or less) an amount, you'd figure, of things done, and said, more or less as in other lands. And with some proportion of misery—for which read 'reality,' if you will wait a minute and not take 'misery' as anything more than a characterization of unrelieved action of words."

—Charles Olson, "Billy the Kid"
Human Universe and Other Essays

Placitas, New Mexico New Year's Day — 1976

[for Ken Irby]

"Stare it well through" but stare it through, following the Jemez
down lava fingers and around Snake Rock, the Rio Grande below
flattens the land out and winds North

and seen that day driving
the "old" highway towards Santa Fe, cottonwood

along the river
branch of the way across

the view of
from Goodell's window

peyote buttons hang from a nail
to charm the neck? worn as an amulet
let each presence pass

friends of good cheer
and wild turkey

all the way across

how many times did you shake down the continent
in an old van? the way we met
"Don't Worry, Be Happy"
hauling your unsettled life across

over each bump and grind
of *open road*
"a land with love in it"
held high over the crown of your head

but stare it through
North and always
it seems West

by what I have come to love
to know the accuracy of
your eye and ear
the music

with the smell of piñon in the air
Goodell out gathering wood

for the warm places
I would like to return
to his wood burning stove
and Fats Waller on the phonograph

sustainers of the spirit
those few friends in that distance

and you are about to turn the ground
from under, again

the heel pivots, I think of your going
North with the present view
to Bridger Pass, then West
high Sierra Pine above Reno
and of course, beyond
the California Coast
"where else?"

as the crow flies . . .
would be the crows in my mother's backyard
cloaked over pork chops
she had thrown out into the snow

two large ones were there first
from the kitchen window I watched the others gather
in the yard and oak limbs
challenging

black wing upon the snow
the burden of being crow ball
and *chain of attention*

The Beauty is the backyard swoop down
from the home ground rises

to the heights of Placitas
not South Chicago, my mother's backyard
and quick melt of an October snow
but John Dowland's "Flow my Tears"
as the van warmed
black wing and mesas

the crow does not return

as I do to the windows
turned away from to rise
to love

"the saddest song . . ." windows and kitchens

again
as the crow flies . . .

"the night's black bird's sad infamy spreads"

took the meat with their talons
and kept the others at a distance

I was thinking of that song
that place

"you can't get to from here"

Variations On A Theme

The interiors,
                        driving South out of Lawrence
mosaic would be the Osage Plains
and the dog leg we made into the West,
                                                the bluestem downns
                                                        [of Kansas
would be the Flint Hills

accumulations
rolling out of Cottonwood Falls
passed the Victorian splendor of John Haskell's courthouse

the back roads
back
across Hollenback's junk windmills

unrelenting wind dervish
Dixie-cups

John Noble's "Morning in the Flint Hills"
layered scenes
from the car window
                        landscapes

⋆

"The day when you could set the points on a pick-up with a feeler gauge in an afternoon with a six-pack of beer under the shade of a cottonwood is gone forever."

over the hills
and brown of day
late summer grass and Smoky Hill River
view from Coronado Heights
Coronado never saw

an initialed heart
carved into stone
then X'd out with a spray gun

as slow as the flow of that river
beyond the recollection of

following endless trails
the heart's sweep, rutted vistas
head on into the Kansas vector
magnitude and direction
as vertical as the whirlwind
up from the SW
weather and geography

lean in

spiral above yucca
blossom above thistle

★

Driving through those hills
Catalpa blossoms along the road
white "showy" corolla clusters
and thread of purple through the petal
the heart shaped leaf I once gave in love
the morning's sunrise viewed from the night before
bare-assed and propped up
legs spread over the brass bed
sweat down the crack of
fucked and rolled in the wash
as though we would fuck no more
the leaf I picked after
on the way back
walking to the house with a sack of milk and eggs

the gift somewhere pressed in the pages
of a Kansas sunset, the sun filtered
over red hills
                    would be the red dust of the Cimarron

not harsh as it was that winter on the Texas panhandle
above the sink holes but the soft tones of the Kansas interior

★

I am engaged with
the landscape's
back pastures
engaged with
what love is
and the *focus*
and misery of
the unrelieved event

the burden of
from car window

the open road
and *the road of excess . . .*

"where that leads"

not to the embrace
but a dark sympathy with
from where I stand apart
the great home of the soul is the open road

★

Twins born in Fredonia
near the Verdigris
conjoined at the heart
of one heart
clutching for the measure

the heart too is
a flower   a vessel

the cup held out in the *mind's eye*
the star both Jung and Whorf
were shown by the Hopi

born at the junction of two roads
and both dead
blur of the heart

buried as twins dress
in twin coffins
or with knees up?

*

off toward some rise
some course
or crossing of attention
from the window
wild carrot behind the barn
Catalpa hedge the land
thin, broken,
purple lines examined
flowers with an orange-yellow streak
on the lower lobe
local grass on the back porch
blown out across the interior thumb and forefinger
rubbing the eyes
at the bridge of the nose
touch the skull
at the temples
from where the hills
roll out

We have lived in more houses than the years
we have lived together

reaching out of

the various fires equal to the moving
the loss of

and now the old Wilson sister's place

sand and lime and hoghair
we live on the outskirts of town
no more than the outskirts of our own lives

a thin layer of dust covers the spice jars on the window ledge
Crocus sharp in the grass
by late February

violet and yellow

along the sidewalk

hold that infinitive "to leap"

after so much winter

Four Fragments Written Over a Passage of Years

for John McVicker

I was standing under an elm near the bandstand
while the band played Dvorák's *New World*

"The negro spiritual *is* the American music."
and then ignored it in his symphony
written in New York City

from those two prominent hills Verrazano sited
turned back to the folk dances of Europe

The band had played *America* first
without the variations
Ives included nor the revelry
of an upright piano
summer evening spilling out into the streets
after the bars have closed

and then the band struck up the brass button marches of
[Sousa
which they played best
and are heard anywhere families spread their blankets

the *New World* came last

"that you is an American"

★

Walking along Coon Creek to McCrary's farm
early summer, Lecompton down the path to the ruins, Fort
[Titus
where Jim Lane sent Captain Walker with cannon, orders to
roust out those drunken bushwhackers
the "grim chieftan" declined John Brown's offer to serve
with his irregulars, but don't deny the voices
freedom cry and rebel yell still ride the ridge of despair
"rolling North to the river"

Below the rough cut of limestone
& down along the wire
on this bright, hostile morning
read the warning
bobbing from a fence post.

Already rotting, a coyote
hung by a nail
driven through his pelvis.

"It's the celebration," McCrary said, pivoting
left hand on the porch railing
and leap over for some more beer

"What does a poet know of death?"
I did not say, except by walking over the ground of the
[dead
"cast a cold eye"

the hills only after so much repetition

★

"I Surrender All"

tombstone inscription   Catholic Cemetery
Fort Scott
                    is not my recollection

but the first time
the loop around Ken's hometown
and seeing two stiff
frozen coyotes thrown off
the bed of a pickup
                                        is lodged between the utterance
                                        and the misery of

surrender

★

By the glow worms Columbus sited the New World
but never understood it

a letter arrived this morning from Copenhagen
poems from Boston
a manuscript from New York City
a letter from Switzerland
and an exchange from California by postcard

all my friends live so far off
this glaciated rim
unsettles even the granite erratic
in the neighbor's front yard
I am that unbudged stone

The Last Deciduous Finger Snaps

Those early daze
schooners wheeled
onto the prairie
the refugees
were so hard
pressed for
wood
they burnt
their axe handles

Verrazano

for David Wilk, a man of the middle coast

by many names

Janus ver. r. azanus	(from his signature)
aka	
Joao Verazano	(from a letter by Joao da Silveira, Portuguese Ambassador to France, to John the III, King of Portugal)
aka	
Giovanni Verrazzano	(Carl Sauer)
Giovanni da Verrazzano	(Bernardo Carli, also, Verrazzano Monument Battery Park NYC)
Giovanni da Verrazzano	(New York Local History Source Leaflets)
aka	
Verrazano	(Henry Murphy and Buckingham Smith)
aka	
John Verrazno	(Francis Parkman)

*

a man
by many
(trans

lations
a map
by his brother's
hand

fragments

all his deeds
and desires
done and
undone

leap
toward that final
abstraction

the land
beyond
the wave's
wave

a coastline
mapped, the story
told
utterly
by each

(re
telling

★

An Inquiry

The year Verrazano sailed was to France a year of disasters,
defeat in Italy, the loss of Milan; while Verrazano wrote
his narrative at Dieppe,
the traitor Bourbon was invading Provence.

If indeed it was his narrative, if indeed he did sight the low
wood a few leagues from Cape Fear, and if indeed he did sail
his caravel, the *Dauphine,* only by day along the coast,
then why did he fail to mention Chesapeake Harbor?
Sauer found this to be a curious omission for a ship
supposed to be seeking a Western passage. Yet, he mentioned no
harbor.
Could he have relied on the Portuguese for such a tale?
This was what Murphy speculated.

And France, without king, without an army or money
and convulsed from within, threatened from without
could hardly, after suffering so much humiliation in a single year,
sponsor another Trans-Atlantic expedition.
It was, as they say, a bad year.

Not much more is known of Verrazano. He was a man of
complex occasions.
After his return, Parkman wrote, he left few traces of his fortunes.
Ramusio stated that on another voyage he was eaten by savages
on a desolate shore in view of his shipmates,
this time sailing under the English flag of Henry.
No, another writer insisted, he was not devoured by savages,
no he was not, Verrazano was hanged
by the Spanish as a pirate at Puerto del Pico in 1527.

★

He sailed from a rock
deserted
close to Madeira

some hesitation
involved
a northern tempest

but he sailed
armed to the teeth
after the repairs

with fifty strong men
he sailed from
more southern waters

Aquarian sky
east-south-east
wind caught the canvas

with sweet lenity
he sailed west
west and holding north

A man with compass
and a fixed star
in his head

his calculations
showed reason
of the mathematic

not Aristotle
one vast sea
but Vespucci

land which isolates
he found no
passage to *Cathay*

He nailed no gold coin
to the mast
as did Columbus

Simply a reconnaissance, listening to
the new world
a matter of time

after twenty-four days
he sighted
a rather low wood

★

He started to return but the waves proved to be too much;
his strength was spent. Exhausted, the waves carried him back.

Seeing this, those on the shore ran to him, lifted him up
by his head, arms and legs and brought him some distance away.
The youth was terror-stricken and uttered loud cries for help.

They placed him on the ground in the sun at the foot of a hill.
They performed great acts of admiration regarding the whiteness
of his flesh.
They examined him from head to foot after stripping off his shirt
and hose.
They built another fire and brought him naked closer to the heat.

His companions in the boat feared, as was the custom in every
case, that the young sailor was to be roasted for food.

Yet, when he regained his strength he made signs
that he wished to return to his shipmates and

they embraced him and watched as he swam back to the boat.

*

The young sailor told his captain
these people were dark
and their flesh more lustrous
possibly greased than the others
they had so far met in the south
they were of medium stature
their faces were sharper
and bodies more delicate
they seemed of less strength
and intelligence than the others

He saw nothing else

★

Arcadia
for the beauty of wood

grapes
as sweet as any
in Italy

Arcadia
for the middle coast

for the vines climbing the trees

for the entanglement
of beauty

this net
of attention

★

In 1508 a ship belonging to Jean Ango brought back seven savages with their canoes and other belongings to the French port of Dieppe. They were thought to have been Beothuks from Newfoundland but the canoes indicate they were perhaps Algonquians.

Verrazano anchored the *Dauphine* and with twenty men made the trek inland through the wood. He found the people had fled out of fear deeper into the forest. Seeking everywhere, he managed to come upon an old woman and a damsel of about twenty years hiding in some tall grass. Both of them had children clinging to their hair. The old one carried a little girl on each shoulder, around her neck was a boy. The young woman carried as many but they were all girls. They were all about eight years old.

As the Europeans approached, they began to cry out. It was the old one who, through signs, told them the others had fled. She was given some food which she accepted with gusto. The younger one was also offered food but she refused everything and threw it on the ground with anger. The boy was taken from the old woman's neck to be carried back to France. They also wished to take along the young woman, who was a delight to their eye, but she yelled bloody rape when they touched her. They had to return through some dense wood being far from the ship and Verrazano decided to release her. They returned with only the boy.

Bernardo Carli mentions the boy arriving with Verrazano at Dieppe in a letter to his father in 1524. But then, he had never seen the boy nor Verrazano for that matter, but by some means had managed to read the narrative. Strange, since Verrazano had addressed the narrative to the King of France, that someone so soon after the *Dauphine's* arrival would be so familiar with the narrative.

★

there were always
the rumors, but Carli's
expectations
were wrong

Verrazano
brought back
no gold
nor drugs
nor aromatic
liquors

for merchants
to arouse
the drowsy pleasures
of court

nor did he even
bring exotic
birds

nor feathers
with the promises
of such birds

no, he
had only kidnapped
a small boy
who he had found
clinging

to an old hag
hiding in the tall
grass

*

Strange, to think
of New York
the city
as two
small, prominent
hills
in the midst of
a great river
broad at the mouth
and deep
within

*

          & from
Rhode Island
apples
          in the margin

(or cherries)

a man with an eye for
the vegetation
oak
and cypress
plums
and hazelnuts

and many
kinds
unknown

to Europe

★

At Newport (?), the inhabitants cared nothing

for trinkets they were given, silk or knives
or iron; nothing at all

even for mirrors
which they looked at, laughed at, then returned their
reflection with a smile

only after seeing the little "objects of fantasy"
small bells and blue crystal glass
did they feel confident enough to board the ship

There were among them two kings
They, Verrazano wrote in his narrative, were the most
beautiful people he had encountered

★

The return
as returns often
are, was uneventful

skirting what must
have been the coast
of Maine
                    where numerous
                        islands
were described
the ship
then changed
sail
headed east
against the northern
mist
          and finally, France

Scenes

(for Carolyn)

Yucca bloom along the margins
of the road

anywhere
their white flowers gently bow above thistle
there is no rest, no shade

travel includes travail
the working of literal landscapes
lateral drifts and measures
we create
scenes, glanced at

held out along the way
wayward

★

(for Jim Schmidt)

The first time I saw the yucca
was when Jim gave them a name
the season after rain
driving an old pick-up
dust and high plains
matched the loneliness of
cracked pods
the seed long since fallen

where love is
born out of that weight
of an endless horizon

Jim drove the back roads
to where he knew of a bar off the roadmap
and a bluff overlooking a sullen little river
which curved north with hardly a care

I gathered shark's teeth
out of the loose earth, out of the sand of a former age

and lost them in California
shaking off the continent
or I would make them a gift
another necklace, another margin
to round out the shape of this vague love

*

(again, for Carolyn)

bent, Dorn says, is a condition
caused by leaning against a bank
as men did in the town where I ate lunch
as they must have leaned on their wheat,
and cattle, oil — Bad Times

crossing the plains in a single day
awaiting the mountains
Pike's first, that great hitchhiker's thumb
thrust out and seen all the way from Limon, Colorado

as a chip
from which you have fucked me down
bared your breasts, as the epic Gilgamesh would have it
with no shame, and I am back in this city
waking from the tangle
fantasy of waterhole
and squeaking bedsprings
rising from the brass bed
light and stone
pleasures of the flesh

scenes gathered out of the air
as if teeth

to go west
is to go against the wind
but *with* the drift is
you said, to go *out,*
to go east is to go back

home? returning
from California

there is no rest
scratching the earth
and the work to be done

*

all day yucca
out the car window, scenes

even Parkman would not trespass
the tenders of the spirit
but walked around an old man's thoughts

cut back through somber pine
until he came to a clearing
and saw the *blue prairie* beyond
which reminded him of an ocean

reborn

beyond

★

unchanged and unsettled
just as obscure as we allow it

this restlessness
O, westron winde
our song seeks
the small rain
the wet of spring
returned

but the dry, cracked pods
of another season
already approach the heart

Vinland Cemetery, Douglas County

These hills do *nourish* fierce attentions
    from tit to cradle, we face the swath made
                                        some imagined terror,
wood cut down the slope, left in heaps
    to be burned, as if this had been the site
                                    of an Indian massacre,
body of wood cleared off for pasture.

The earliest settlers, Malin guessed, were Mennonite
    gone under to the grave
                                        or the oldest land grant
library in the state,
                                now a line of crude houses
                                        a church boarded up,
just a slow place.

                                And a December wind whips
out of the North across these unpretentious stones.

        Irby follows the German names and family plots
                                            to the wire, a line *flat out,*
again, these hills.

                                This is the only of all possible worlds
where it leads to a slow place along the Santa Fe Trail,
                                        marked only by a few
eroded passages.

                                    Death angels do not smile
                                            from these carved stones,
these idle days, these hills.

For Hart Crane

"The source of the song must die away"
Robert Duncan

★

after so much troubled
the sea is
she is the flower
accepts our dreams of moon

"who is the woman with us in the dawn"

★

The hand of fire
gathers wood, prepares a flame which will
[burn all night
"What was that line carried so long and now forgotten?"
set in moonlight, forgotten by dawn
each memory with its own stray rhythm
as sharp in the air as this morning's crushed fennel

various night fires
back to those moments before the kindling catches
each betrayal, each agony of a kiss

"Here at the water's edge the hands drop memory"
black willow along the river
how long ago was that?
hands dirty by all their made gestures

rage beyond meadow
Hart shouted as Lear shouted No

★

The hand of fire drops a coin in the jukebox
and it would be an old line
from mid-kingdom
King Oliver recording with "Red" Allen
"Stingaree Blues" 1930
the year THE BRIDGE was published

★

The last sawdust of Whitman's America
settles on a midwestern roadhouse floor
saxophones and stardust, shotguns
"from Kill Devil's Hill"
                                        and Wright brothers flight
straight to that golf game on the moon
viewed from the vacancy of asphalt and concrete
Chicago south suburbs
                                  not *our* moon
nor lunacy
                but like the woman from Atlanta
asked the reporter, how come
the broadcasts from the moon were clearer
than the soap operas? meaning
it had all been done with mirrors
she was, simply, handing hers back

as did the Nanticoke women
returned the mirrors with a smile to Verrazano
                                                                summer 1523

*

Binford's distance boredom is measured by
a cigarette pack thrown out the window of a passing car
or beer cans dropped between hunter's legs

why was I thinking of that assemblage
after your line, Poe's "retching flesh"
shakes and delirium, again
hands more than gesture, a voice
his last night through Baltimore
"Did you deny the ticket, Poe?"

except to take the toll of such crossings love is
a terrible, hopeless thing

to be in the service of
we are made to suffer

so I wanted to know the music at my birth
the dance of my mother's rotten ovaries after cancer
feeling a kinship
meaning kindness
too quick walled up

★

What were you listening to in Cuba winter 1916?
how much of Africa did you hear
and return with
where the soul crosses
the bridge from your window
the *unknown* lover's
fingers locked in yours?

Failures, too, go on to the end
Havana nights
          my mother
detained at the hotel after Castro's revolution
while I finished my first year at military school
my first poem
          roses and angels, the Japanese tank
          we painted flowers on
          under the foliage of a spring night

          not knowing of the thorn

and where could you go after the wake
of sea
    but to the wake of the dream
and from the stern of the *Orzaba*
looking back toward Azteca
dark and hypnotic
           her legs up into the black obsidian blade

how quietly did you fold your coat
before you leapt?

(this is no requiem
nor meant to raise the undiscovered dead)

★

Taken by the collar and led out of the party
peeing in the dark, standing with legs spread
dogs howl by the river
"you'll wake the dead with that howling"
a friend said who is now dead

I am believer of
                    the watchers
                    those keepers
of flame
          those dear ones
the hand of fire reaches out to

or what is left after the last ember cools
but to take me

as much as I hold the crow to
my carrion

I hold more the last keeper
source for meadow and goats
gods and the text of the fleece

or we stand so far and alone
then
       leave me there with the dogs
and you to your sea

A Past Winter Recollected

Awake those vast moments before dawn
in the kitchen she moved through wearing a green housecoat
making coffee for us and breakfast for the children
then back upstairs in the first light to get them dressed
before she dressed and down again
after I have poured two cups and returned to the table

and in that light I remember the outline of her body
which was not there through the housecoat
but in the memory of her while she stirred the cereal
on the stove and then took her cup
first taste of coffee standing
in the body of light through the domestic
more so in the lines of the body of words in the notebook

where the day comes forth cold, clear February
last year's crocus during Duncan's visit
the page turned back through pleasures
now removed by winter
instead, these interstices
broken, shattered places of ourselves
and how *the heart* was revised out from the lines above

*

The mind more in seasons angled into winter
my own lack of a grasp more permanent
returns to the pit I dropped the light and heater
to keep the pump from freezing
and the frog awakened
with a few green sprigs of vegetation
wet, even balmy
"plunk your magic twanger . . ."
from the great, dark reaches of the pond beneath the soil
through "breaking the silence"
doing the digger things we do
late for work, o shit, Christopher late for school again
or tacking back the sections of the roof blown off
during the storm, caulking
checking the pump

⋆

forsythia and crocus, for love
but not yet spring lush vegetation
nor out of this morning's wood
almost translucent through the window postcard
Frederick Church's "Jerusalem from the Mount of Olives"
on the ledge where travelers meet
and exchange the event of

at a distance from the city
bathed and patched by shafted light
meadows where angels dance between the clouds

and at that distance the light I remember
has always been artificial driving at night
a city where Gilgamesh lords it over
and so many after Enkidu have come down
each from their temple
for me it was Los Angeles from the San Bernadino Heights

but this morning opens with olives
forever dark in the foreground
the light unfocused but what I have gone by
postcard narratives and the accuracy of others
overland with beer in the cooler
maps and white crosses on the dashboard

★

Hart Crane's merging morning seed
"The last bear, shot drinking in the Dakotas"
given Pound's theory on the minting of coins
and Roy's dismissal, as long as there are nickels
there'll be buffalo,

there is no Jerusalem beyond the ledge

*L*OVE *P*OEMS

Some of these poems may seem fragmented and at times obscure, but still they are suitable for everyday use.

". . . how what we ride rides us."
Charles Olson

"Everybody's looking for a bell to ring
all through the night."
Lou Reed

Beginning With A Line By Robert Duncan

"Let grief have its voice"
love fills the cup
the loss the heart would endure again
I stood on the front porch listening
to the various wind chimes down the street
all the way to the river
New Year from under the storm
and halted celebration seeking
something other not just someplace
somewhere else called up
but stood there as dancers stray
too far from the music
suddenly there and brought back
to the falling snow
trying to remember what the view
from the porch of each house I have lived in
would look like at this moment
where each Russian olive was planted
Oaks and the single cottonwood
of memory

"there is no end to desire"

spent the night spent arguing for a place
from which to view my own eternity

I deny dim memories love and war attend
but begin with an urgency beyond desire

cottonwood denial
still shimmers in the leaves
of recorded years

other than *knowing*
thrown to
affections
a well of hyacinth
others dipped into
now your fragrance
on the sheets
lingers
long after the memory of you being there
a rose would be
a rose of
                Sharon

tonight a red sky rides out a local anguish
across the sting of scorpio's tongue
taking our clothes off
and jumping into the swimming pool
into the quickness of
mars in your eyes when I kissed you
caught in the silver net of surprise

from the first house
uttered love
to the next
for the first time
your thighs to the next
renewed by
I said after
said again
until we both made that pledge
pledging love
as if we had taken a dare
in order to be free of it

a shag of wheat
should we last so long
the unknown *high* of
your eyes
green I've been touched by
and want to touch you
into winter
making love made

the *arched* you the line dictates
a shape beyond duration
maybe mileage
or at least a roadmap
taken across you

a stone hurled over corn
to be as tight with

somewhat retuned
it wouldn't do
to say I love you
so from another source
out of sight
into the heart's play
I miss those twins
of Fredonia
cojoined at the heart
but not of a single mind
reread *OREXIS*
opalescent glass
broken
dedicates love
to be as obscure as we allow it
somewhat returned

let the *bay* you *be* my laurel

your eyes are no harbor
but reflect the parrots and monkeys
the spices of the orient
in the stalls along the Ramblas
your eyes are the market place
until they flash across the sea
and there's a storm
and the whole country closes shop
your eyes are not exactly the music
but the beat and shake of a tambourine

a poem tapped out on your back
so softly as to be a rune

don't give to blue the scent of memory
but from silver the cup overflows
the day after equals silver
doubts and sad mirrors
reflect broken figures
of a slow dance mined in the underworld

Red birds
Pound's opera
in the red bud
blooms
in cadence with
the storm
and stress
of another Sunday
torn from the world's
string

O my venomous beauty
a scorpion hangs from each ear
out of silver
                                        on your birthday
for you belong to the dark, unresolved south
when you turn away there is silence
and the territory your eyes scan
is certainly, unexplored

and instead of saying, come back
I recite the *rose* from a journey
I unsheathe for you out of season

and have been stung by your bleeding

the angels have left me over this
I must find my own speech for wine and tobacco

There is no other landscape
than the one which fills my rutted mind
I shall not compare you to hills we trespass
though the footnote to this passage
might mention *to nourish*
for I have not strayed too far off the rim
from where we sat on the full tilt boogie
of the Gridleys' roof toasting the solstice with them
in the day's last length of light
filtered through Roy's dreaded honey locust
O my scorpion, to be with you and friends
on a day so sharp as this
is all I want of hills and wood and light

"Venice,"
President Grant told
the Ambassador,
(according to Henry Adams)
"would be a nice town,
if only they'd drain the canals."

I wanted to find a poem I had written to read to you on Poe's birthday, but it was not where I thought it to be. So I rummaged the shelves for an hour, drawn at times to other poems, and when I finally came across what I was seeking, the poem turned out to be other than what I remembered writing.

May thought be
inside a thought
I have touched you by
and echo in a line
even though my left handedness
has slurred the page
or there would be a meadow
when there was a sea before us
and finally I could not read it
and now set that spring
again to flame
a tide withdraws
taking some small part of us
along these restless avenues
where only lovers would consult
a moon as pale and deceptive
as the one above us

No horns other than the dong of carnival
Venice is a quiet sinking splendor
even dogs go muzzled through the streets

Burn the artist but save the canvas
pictures, churches and ruins
we attend Veronese's *Last Supper*
standing with Christ's jugglers
closer to the wine than the table
but I have told the inquisitors this
I have told them of the heart-shaped map
the still undiscovered imago mundi
and have burned it in my thoughts

while we walked aimlessly
lost in the maze of
narrow winding streets
lovers walk in idleness

pigeons eat from the new world in my hand
under the shadow of a church and the doge
three days without a printed word
I write no more postcards

Lost count fucking
under the weight of
the waves' wave

my eyes roll back
into the skull of
love's skirmish
and unroll
an old map

the heart-shaped one
of the 16th century by Fineaus

for those deep seas
lost sight of
the coast

you make it so.

Too, for Aviation

1

On the 20th anniversary of
Moon Landing
still there are doubts about
those alleged astronauts,
their best defense was
they used it
as an opportunity to play golf.

But I remember what the woman
from Atlanta asked the reporter at the time
how could the tv reception from the moon
be clearer than her local soaps?

as if it had all been done with mirrors
& she was simply handing hers back
like the Nanticoke women returned the mirrors
to Verrazano with a smile in the summer of 1523.

2

The stealth bomber lifted itself
for the first time today, circled gotham
then returned to ubiquitous applause

all this accomplished on the 50th anniversary
of *wrong way's landing*
in Ireland, so

this day of stealth & Corrigan
is to aviation what the *New World* was to Columbus.

The Latest Petrol Crusade

Labor Day, 1990

From his lie in the sandtrap
Bush said, "This isn't going to be
another Vietnam," & waved a 9 iron
straight at Baghdad & Saddam headed
for his german made bunker
and read poetry

The Golf was played with precision
& yellow ribbons & stars lit up that night on CNN
like *the last invasion of Christ*
except for a few oil soaked feathers
& like they say, *friendly fire*
we can take pride in our leisure.

Overheard

at the diner a denizen theorized
that if China went to a market economy
& *everybody* bought a frigidaire
by the 21st century there'd be a hole in the ozone
the size you could drive a truck through
& that truck would be the size of the North American
[Continent
with a camel on the side "Humpin' to Please"

Mercy, Mercy

forget Pound's tack hammer, Jack
Disney could redi-mix a three day
town like Florence
& recreate it in Omaha
quicker than you can say
Kulture!

Poem for the NEA

"Holy asbestos, Matban,
the American Flag Co. is having a fire sale.
Can you grant that?"

"Mapplethorpe nothing," the congressman said,
"until I've unpacked from Barbados."

The Allegory

So kill your father
get nine more
railing in the big oak limbs
& walk out from under
all that shade

your mother is your

language.

Cahokia Mounds

We ascended the great mound at midday, climbed Monk's
[sweating
distance & duration hinged to celebrations
to view the flood, fish heads & then corn w/a touch of
scandal/sacrifice where the two copper wands were found
& bones of young girls waved over postcards
make narrative of penetration, but our view was corn
& the St. Louis Arch through the industrial haze, a race track
& across the street an X-rated drive-in *Now Showing* . . .

Reunion

There is a sadness which inhabits reunion
in the old songs no one quite remembers the words to
& the lost embrace long after the campfire has burnt out
where once there was a flame I find myself
going through the smoke

There is a weariness to reunion
after all the tablecloths have been stained
& the talk of *for love* has used
the lodge poles to rekindle it

Postcard

for Carolyn
Anon-Hell-ala Hudson

you are so
ordinary antlers
nailed over a mailbox
never fail to remind me of
you!

Lauren Hutton by Richard Avedon, Grt. Exuma, Bahamas

Postcard

for Margaret
in Moscow
via Amsterdam

Cunt strummed cicada nights in the backyard
where once a yucca bloomed, an echo.
Big mama whore you were all Babylon for me
those years before
the last invasion of Christ!

Carhenge, Alliance, Neb

Postcard

for Bob
Idaho
        &
out

like you say re: Anne Waldman's reading,
she made us eat the headlines
of greed & dysfunction
with the voice of a noisy neighbor

but, at least, she's one of those gin drinking,
chain smoking, Naropa Buddhists, isn't she?

Certainly, she wears a great scarf, right?

Joseph Cornell, A Pantry Ballet (for Jacques Offenbach)

"In fact the debasement of
coin really begins when the
design ceased to be cared for."
Ezra Pound

"As long as there are nickels
there'll be buffalo."
Roy Gridley

the sun goes down while heaven waits
& Sappho's moon shines brighter than a flashy coin

Devonne's enigmatic smile could redeem this world
like a flood, if stars were nickels to our thoughts

O Roy, if stars were nickels to our thoughts
she'd drop them in her pocket

"I'd be playing this *shit* even if no one was listening."
Charlie Parker

I snatched the sun's eternal ray
& wrote till earth was but a name

your pledge as one crazy poet to another

O Clare, badger I have been likened to by Beth
slow to the ground & with little intelligence

& the days walking in Essex wood are so few
we're treed & the dogs are howling

they ain't heard nothing yet!

badger me & badger wood
while waiting
badger you
                    while waiting
in the wood

Hot Nights in Rio

Love is eternal
as long as it lasts

They say as they say, they're about to roll out
those dreamland floats onto the colonial streets
winding toward the boulevard

while forsythia blooms
my roto tiller, so early in the backyard
of the left hemisphere, back streets make of
privilege

& they dance who have no other reason
certainly, nothing to dance for

while I sleep, while I sleep Rio is a frenzy
& when I dance Rio sleeps it off

soon the redbud will bloom

I have danced & slept & now refuse

So tonight might as well be carnival
with you running ahead of all the rest

tonight might as well be a moon as cold
& hard as a barite rose.

On Whether or Not
The Pig is Indigenous to the New World

from *I Hear America Cooking*

My own theory is that the pig landed near Miami circa 1542
and left the same day in a rented DeSoto.
Sometime later, the pig arrived in the SW
and enjoyed the distinction of being
the only adventurer of European origin
to be both domesticated and savoured in that region.

In Greenland the Inuit use jackhammers
in order to break through the ice
and bury their dead

my own climate has the morality of a chainsaw

"The white man will freeze to death in the arctic,
if left to himself," the elder said
after translation, "but now with schooling
some Inuit too have frozen."

How would you feel

if your dog ran off

with your imaginary friend?

Snapped in River City

lee chapman ©1988

Pastoral

Hitchcock Co., Neb.   24 December 1993

No matter how much you wish for
Ella singing Cole Porter, it's easy to see why
they call the substance of this place massacre canyon
with all them Sioux rioting in those days

but what do you do when Miles hits the deck
on a vast, open stretch of acceleration
with a bye, bye sassy blackbird
                                        & one of those real
ambitious sunsets, for all the world a ball of petrojelly
called upon to light a cigarette, rushes parallel
for awhile, then drops like big bird over Lockerbie
behind sand hills?
                        bye, bye!

Precambrian Slab

unravel travail & you get Bolero spewing out a mobius ribbon
   from the tapedeck
on the high plains of a slow confluence when yesterday was the heart
   of a subterranean turbine on the *industrial walk*
now even lunch is difficult off the afternoon roll Junction City
   makes of Seoul food, Asian whores & shopkeepers who have followed
   the tents to Ft. Riley
& tolerance is a Brinkley goat gland transplant to *keep it up* & swinging
across the radio foam promise to have the whole world *in your hands*
   with Reichian joy
going east means make it quick & uneventful
after coming down from the foothills last night drinking coffee & cognac
   with chocolates served on a silver tray
back to the mosaic of cheeseburger, home & beer, *the elms are dying*
on the way irrigation circles are jerking corn west of Salina
where the testament is in full form, *go for the loot!* as big as the bottom
   line on a billboard
while BBQ Bob plays the washboard blues from a Boulder rooftop

what this country really needs is another ice age

Go for the

6/July/92

Rave-On-Fog-Woman-Spawning-Salmon
pole narratives everywhere rot while we walk
through the mist uphill to the clarity
of Miles Davis on a boom-box
while two Tlingits carve red cedar with a chainsaw
somewhere women weave a chilkat blanket
goat hair around a single strand of cedar
cedar in the air as thick as the mist
low slung this morning from the narrows
where we watched two ravens badger a bald eagle
off a wharf piling for our amusement
such theatrics not unknown to the Northwest
stagecraft as far South as Kwakiutl
open-mouth Raven Cannibal Society
Boas, who wanted to set the record straight about "primitive"
posed native for the Smithsonian, squatting
anguish before the leap through the blow hole
tricksters like the bird all the way up the inside passage
a tear in the mind a world away, sadly to say, Ketchikan
was a four day town, and we ferried away
with only a sense of where the penny drops
on an island torn from, even floating
where the earth is saturated with arsenic
no gardens, vegetables flown in, celery
celery and poached salmon, and a sawmill spewing pulp
soon to be ice cream, O Sharon, the feast is redundant
let's make our wedding band those ringing, inverted
conical foxglove bells.

Ingersoll Hotel

7/July/92 8 am

Raven had no commerce with, nor use for
was constantly annoyed by
so one day did a little conjuring
and turned all of us into the hides
we wore, only those who went naked

remained human.

Turning from the narrative
to a cruise ship blocking out
all but the fog, watching that very
slow process of those disembarking
with a certain detachment one has
from the privilege of a hotel window.

> *To live in Alaska*
> one must be resourceful
> and a bit renegade.

Saxman Totem Park

7/July/93

Totems can relate the lineage
of a person or visually narrate
an event whether it be a move
or the creation of the sun
or moon or the exploits of a brother
or a lover or the foundations any poet works with
so of course, there is a pole
reserved for ridicule
or honor, lean with a head
at the top, Seward went to Alaska
and by all accounts the Tlingits
found him ill-mannered and parsimonious
and carved a potlatch hat on his head
because he interfered with
and then there is Lincoln
standing by, stove pipe
how his regulars stopped a war
is honored, kindly, grim
Lincoln, the woman at the Native
Artists' co-op told another
take, by the Emancipation Proclamation
he too interfered, well, the town gossip
can also be a head for ridicule.

Mud Bight Totems

8/July/92

"They have done so much
for these people," said the Greyline
fishmonger about the 19th c.
Protestant Missionaries.

"You mean the Rev. Duncan who locked
children in a closet whenever he heard
them speaking their own tongue?"

"They were all like that," she said
walking away.

First alcohol and syphilis and then
the missionaries . . . I wanted to say
but said, "Not the Russian Orthodox.
They kept their distance and translated
the bible into Tlingit."

But it really didn't matter
for either of us
the whole cruise ship
had piled into the bus
and wanted to go shopping
in Ketchikan, buy those cute
little ersatz totems at Tom Sawyer's
what the Tlingits have been mass producing
ever since Baranoff.

Flint Hills, Limestone Rocks, [the Cradle]

Thin mosaic west of Council Grove, last chance
hardwood memories for the spokes of commerce roll
the solidly built Madonna, the *solid sod*, plains memorial
children clutching at her skirt
the way to madness by their look,
sex and drugs and Dvorák's *New World*
spun out during the storm the year we had to be towed and
  missed
the rant at the party in Lawrence

not even an indigenous Sunflower but seeds long since fallen
off the mud caked wheel's rutted track returning from Santa Fe

rained from Overbrook autumn in the air between the pages
our constellations celebrate this passage
slowing down at the stump-signed treaty with the Osage
paper not fit for wipe where the band plays Dvorák
when it's humid, post oak shade for a letter drop
where Custer camped by a roadside marker
or a Holliday tavern, was it, a Holliday driver, Cannonball
a Slim Pickens of a man ordered Carrie Nation out of his cigar
  smoking coach
she walked with hatchet for a Noble nude, hacking
Cleopatra at the Bath during a raid on the Carey Hotel
a gross and simple obsession became her trademark.

*

mock organges fall, bois-d'arc, Osage childhood bow of
gypsy gleaners, prairie fruit for what purpose other than
grandmother's centerpiece, oak leaves and acorns
at Halloween, treats, while the mind strays driving
focuses on an abandoned farm house
or a *few scrub cottonwood, "as dancers stray too far from . . ."*

the strata of former drives, notebook lines
X'd out like with a spray gun over the initialed hearts
etched in sandstone at the WPA castle, Coronado Heights
a blade of bluestem so deeply rooted to the nub of *going west,*

Going away . . . a former knapsack with *On The Road*
like Parkman's Byron in his saddle bags
or Milton's *Paradise Lost* companion
for Garrard on the Taos Trail
Better to reign in hell than serve in Heaven

[toto burn out]

Oh Westron Winde, we
done sprung
over that
lovely, lonely, fucking *rainbow*
and we're off
to see the yellow
brick lizard
with road songs
rhymed
for the drive
titles off the wheels of
our spinning Wurlitzer
like, "Olde Jack Black Kerouac
was thumbing for a place
to be a scene,
                                        zippity."

under an egg crate ceiling
after a bierock
south of the fertilizer plant
and between the freight
Sharon, can't you hear
the nitrates stirring?

★

A narrative written over all past narratives
recounting *the crossing*

to be an artist
who stretches
a canvas
with an eye on
the landscape

what Noble did with a diaphanous *Morning in the Flint Hills*
accentuates the other layers, creating
hills with a palette knife

. . . that morning after the storm, north of Marquette
clouds breaking as form does to illuminate

a landscape, for a moment
transfixed by a shaft of sunlight, impales

the random.

Dedication

for Sharon

Don't you just love these days
A Mu sings,
"take my soul but let me linger"
where the pine shades the foo dog rub
and the weather changes, a spate of crows
maybe circle above us, or reminds me
of Dante's thrashing whirlwind *Inferno*
O my a turbulent sky to Amuse us

POOR CECILE

Poor Cecile

for John Gary Brown

surrender all, *Poor Cecile,* those untold days
of thunder or drought, but a coarse
granite touch, all there is
that is glacial, marks you

as the crown of earth is
so light before a wind
and love all too soon kindles
a gentle heart
so quick to take to flame
as we imagine Beatrice
and Dante stalking her
the distance we see in you Cecile
no greater grief from ashes
remembered than to remember
those brief days of joy
when misery is at hand.

Retuned

from the Popol Vuh

Out of a watery chaos, null and
void, the ancient gods, Tepeu and
Gucumatz fashioned the earth and
grew bored. These divine generators
realized plants and
animals were not enough to venerate
their ingenuity. So they shaped
proto creatures from mud but
all too soon to mud returned.
A race of stick figures
were next to make the scene but
these manikins couldn't even
keep a beat and were torched, only
to be replaced by men and
women made from flesh
who had the rhythm but
were carnal and virulent.
These forefather gods anxious for
praise called forth a black rain
to flood the earth and annihilate them all.

Tepeu and Gucumatz, who
created the earth and
endowed it with plants and
animals, started all over again.
This time a true race, kindred
to the gods, was kneaded from
a dough of *kayem*, Maya
for ground maize.

And the gods sat back.

Abrazos for Michael Coe's *Maya*
from which this version was retuned.

Delancy's Bar & Grill

Grant, Nebraska

We come in out of a rain
that could wet a sidewalk
but not quench a thirst
and our bartender asks
"You aren't one of them
goddamnecologists?"
run on just like that mr cummings
So I lift a line from Lou Reed,
"Oh no, we believe in dumping
battery acid in the stream like everybody
else," rotating away
from the counter to watch
the kids shooting snooker,
a class act, unlike 8 ball slop

"What do you want?" he asks
utterly without malice

"Whatever's on tap and not lite,"
when I realize our tender
could go on Oprah
as the spittin' image
for a Jimmy Carter

"What are you doing in Grant?"

"Visiting her family."

He turns to Sharon, "Who are?"

"The Hoffmans."

"Oh yes, he has that restored Chevy pickup.
See him driving around town
all the time, though he's never been in here."

"That's okay," I say, "we'll make up for him."

"There Are No Bridges in Perkins County"

for Samuel Cassidy

There is a hi-way marker
where once there was an artesian well
"Wild Horse Springs" some distance away
where your great-grandfather hauled water from
the county line as a boy
by wagon back to the farm west of Grant
until Ford's assembly Model T
reached the high plains
and Darryl's been a driver
since

"Amaranth!" your great-grandmother
said, turned to the back seat,
for cosmetics,
dark, reddish field, drought resistant
sprung from the stem of waste, wither
and quench, imaginary
I turned to the west, sunflowers
Blakean bent tramps
weary of their large, domestic heads
while we drove north to Ogallala
then east to Ole's for lunch

*

distance
                    as a stretch of history
told from the depressions
where the house once
stood
or a barn
shed and sink hole
beyond rutted gravel
the one-room school
house your
grandmother
                                attended
the drive was
out to the farm

*

Your great-grandfather's brothers went off to war
while Darryl stayed and farmed the land
before that he drove a rig
from Ogallala to Omaha
courting Alma whose parents disapproved
for the sandhills were then as now
a backdrop for unabridged prejudice
they were Mennonites, furniture
store proprietors,
the Friesens were all propriety
and cause for dismay

                                        when, Sharon, your grandmother was born
five days before absolute respectability
shotguns for birds and weddings

*

upon his return from the Pacific rim
the mother gave
the brother the farm,
so Darryl moved around a lot
working slow places, your
friendly butcher
or hauling, with a mind's eye for
and never strayed far
from the platte bottoms

helped raise Great
Aunt Billie's   kids
more or
            less
                  and
not a felon in the bag

though behind the facade of family album
there remains a bitterness
"He didn't want the farm," Darryl
said, following me out of the house
while I lit a cigarette, "and didn't know how
to work it."

                  "Gosh, John, I could have
made something . . . "

Darryl Hoffman
[10 February 1912 - 28 December 1998]

~ *Afterword* ~

IN THE SPRING of 1964 I walked through the sally port at Culver Military Academy turning my back on a life determined since adoption. But action isn't always focus. After a short stint in northwest Missouri, I lived with my grandmother in Pekin, Illinois, until she died. Her father was a coal miner. Her husband had worked his way up to the high rungs of the ladder at Keystone Steel and Wire. Though I was his namesake, he referred to me as the Sears and Roebuck kid until his death. My grandmother was kind and for the first time I had, like they say, space. During this period I purchased at a bookstore on Well's Street in Chicago a copy of the Donald Allen anthology, *The New American Poetry*. Who didn't? As outrageous as the title seems today, it was accurate then. I used it as a guidebook. While living in a boarding house in Peoria, I read an article, I believe it was in *The Evergreen Review*, though it's just as likely to have been in *Life*, which was always on the coffee table, about Allen Ginsberg's and the Fugs' romp at the University of Kansas in 1965. I applied there and was accepted by return postcard.

I first met Ed Dorn on the Hotel Eldridge roof in the spring of 1968. We talked about Colridge and then our attention moved to the Kaw bottoms, to the willow and cottonwood along the river. That was a singular moment, still no more than a para second away. Dorn suggested I read Carl Sauer. The following spring, when I wasn't ingesting "spiritual condiments" or throwing rocks at the ROTC building, I was attending Dorn's poetry class. Robert Creeley and Robert Duncan were the three-week guest lecturers. The talk took the form of collage. Place was no longer linear but composed of layers and levels of pursuit. They were the Dons of the Black Mountain Mafia. This book springs from that experience. I love these poems and where they went so often on their own course. As a poet, I see myself in the backseat voicing my concerns about the driver while keeping an eye out the window observing landscapes, whether they be the stark abundance of the Flint Hills or the canals of Venice, observing the fortunes of greed or love and the miseries.

John Moritz
July 2001
Lawrence, Kansas

Other books from First Intensity Press:

Tokinish, James Thomas Stevens (poetry) — $25
[hand-sewn, signed & lettered by the author]
The Man in Green, Patrick Doud (poetry) — $10
Kicking The Feather, Duncan McNaughton (poetry) — $10
Flaubert At Key West, Barry Gifford (new & selected poems) — $12
Perturbation, My Sister, Kristin Prevallet (prose poetry) — $10
The Roads Around Jenkins, Theodore Enslin (poetry) — $10
STUDIES: cuts shots takes, Kenneth Irby (poetry) — $12
FIRST INTENSITY (annual literary journal) — $14
[back issues $7 ea.]

For information, write to:

First Intensity
Lee Chapman, editor
P.O. Box 665
Lawrence, Kansas 66044
email: leechapman@aol.com

~ ★ ~

First Intensity publications are distributed by
Small Press Distribution
1341 Seventh St.
Berkeley, CA 94710-1403
email: orders@spdbooks.org